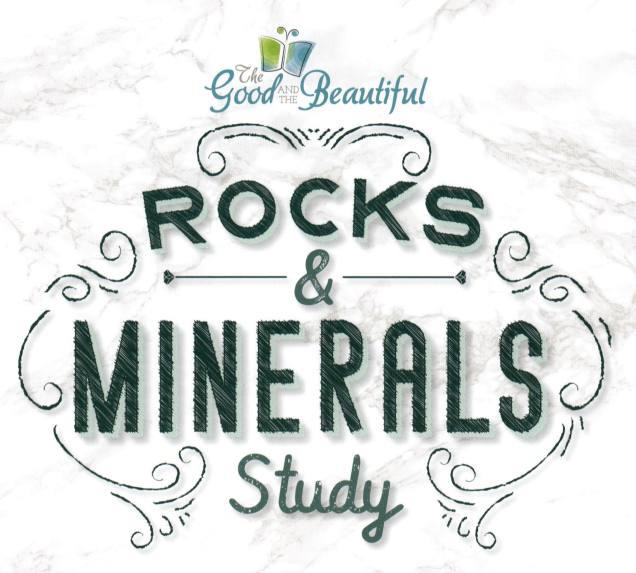

DEFINITIONS

CABOCHON
A stone or gem that has been smoothed and polished, but is not faceted

FACETED
A mineral cut to have many sides, increasing its apparent luster

IRIDESCENT
Creates rainbow-like colors when looked at from different angles

CLARITY
How clear a mineral is or the ability of light to go through it; translucent, transparent, and opaque refer to clarity

MOHS SCALE
Rates the hardness of a rock or mineral

OPAQUE
Light does not pass through it at all

TRANSLUCENT
Light goes through it, but is foggy looking

TRANSPARENT
See-through, like a clean window

LUSTER

The property of a mineral, crystal, or rock that characterizes how it reflects light

ADAMANTINE
Extraordinary sparkle that appears transparent or translucent; minerals that are not quite as sparkly may be classified as subadamantine

VITREOUS
Looks like glass, but without a lot of sparkle; appears to be transparent or translucent

SUBVITREOUS
Glass-like

PEARLY
Has a luster similar to a pearl; somewhat iridescent

METALLIC
Opaque with surfaces that shine like polished metal; minerals that have less luster, but still look metallic, may be described as submetallic

GREASY
Looks as if it was dipped in oil; may even feel oily or greasy to the touch

RESINOUS
Has a plastic-looking appearance

SILKY
Appears to have fine fibers, all running the same direction within the mineral

DULL
Sometimes called "earthy"; has very little to no luster

WAXY
Has the appearance of wax

Scolecite

Scolecite is an exotic mineral that grows inside small cavities of volcanic lava rock. It is usually a spray of prism-shaped needles, most often clear or white, but it can also be salmon, pink, green, or red. The needles can be very brittle, so they must be handled gently.

Rock Study

Doesn't this mineral look like it fell right out of heaven as a snowflake? Imagine discovering it in the cavity of a black lava rock. Can you picture the contrast of the crude black lava rock compared to the brittle white spray of scolecite needles?

Facts

TYPE: Mineral

MOHS SCALE: 5–5.5

LUSTER:
- Vitreous
- Silky when fibrous

OPAL

FACTS

TYPE: Mineraloid

MOHS SCALE: 5.5–6

LUSTER:
- Subvitreous to waxy

Opals are either precious, common, or synthetic. Precious opals have a play of color in them that common opals don't, meaning you can see reflections of colorful light inside a precious opal. Synthetic opals also have the color, but there is a way to tell them apart: precious opals are fluorescent under a black light, and synthetics are not.

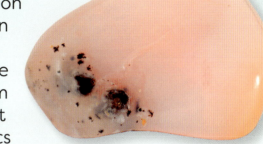

Natural blue iridescent opal

ROCK STUDY

Look at the three stones of precious opal. Gazing into them, doesn't it seem a bit like looking into outer space, with different matter floating around? Count the different colors you see inside the opals.

Pyrite

This lustrous metallic mineral is nicknamed "fool's gold." You can see why! Pyrite is used in a number of modern industries, including jewelry, paper production, and solar panels. It gets its name from the Greek word "pyr," which means fire, because it sparks when struck by metal.

FACTS

TYPE: Mineral

MOHS SCALE: 6–6.5

LUSTER:
- Metallic

Rock Study

Do you think you could be fooled to think it is gold? Here's how to tell the difference: Pyrite is hard and brittle, and gold is soft and bendable. Pyrite usually forms in cubes or crystals, and gold forms in random shapes. Pyrite often has striations, or visible lines, on it; gold does not. Can you see the difference?

Pyrite striations

LAZURITE

This semi-precious stone is mined mostly in Afghanistan. Bright blue in color, it usually has stripes of minerals running through it, such as pyrite and calcite (which is white). In historical times, it was ground into a powder and added to oil to make paint—an extremely rare and expensive blue paint. The stone is often used to make jewelry.

ROCK STUDY

Doesn't the beautiful round lapis lazuli (lazurite mixed with other minerals) resemble the earth, complete with calcite clouds and pyrite islands? Look closely at all the specimens pictured: the sparkle of the gold pyrite, the darker blue inclusions, the white calcite. Do you think lapis lazuli would be as beautiful without all those "imperfections"?

FACTS

TYPE: Metamorphic

MOHS SCALE: 5–5.5

LUSTER:
- Dull
- Vitreous
- Greasy
- Waxy

Cyanotrichite

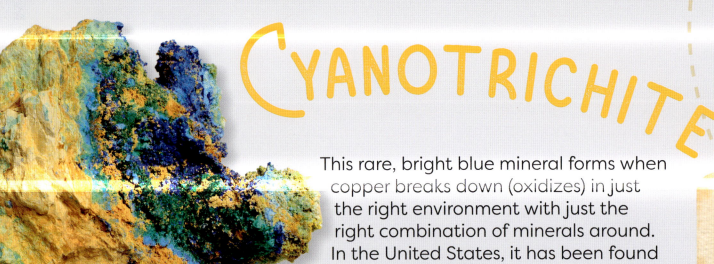

This rare, bright blue mineral forms when copper breaks down (oxidizes) in just the right environment with just the right combination of minerals around. In the United States, it has been found in Utah, Idaho, Arizona, and in a mine in the Grand Canyon. Its name comes from the Greek words for "blue" and "hair."

FACTS

TYPE: Mineral

MOHS SCALE: 1–3

LUSTER:
- Silky

ROCK STUDY

Imagine holding a "blue hair" ball of cyanotrichite in your hand. Do you think you could be gentle enough that none of the tiny, brittle needles would break off? If they did break off, how do you think they would look in your hand?

Agate

Agates form inside other rocks, such as lava rock, and take an astoundingly long time to form! Agate is a type of quartz that has bands that start on the outside edge and move inward. Agates are formed in all kinds of patterns and colors.

ROCK STUDY

The bands of agates often create very interesting patterns. Use a piece of paper and pencil or markers to draw an agate with bands. Make any outer shape you want, and then continue drawing bands inside to create your own unique agate.

FACTS

TYPE: Mineral

MOHS SCALE: 7

LUSTER:
- Vitreous

Marble

Most people know that marble, often with swirling, colored mineral veins, is used in sculptures, countertops, and floors. But did you know that marble is also an effective acid neutralizer? Have you ever seen people taking antacid tablets? Those have ground up, processed marble in them!

FACTS

TYPE: Metamorphic

MOHS SCALE: 3

LUSTER:
- Dull
- Pearly
- Subvitreous

Ground Marble

ROCK STUDY

Look at each of the pictures of marble. Notice the multiple veins of different colors that run through them. What do they make you think of when you look at them? Do they look like lightning bolts bursting across an ocean scene? Everyone may see something completely different. What stories do these pieces of marble tell you?

Calcite

Calcite grows in all kinds of colors and shapes. What distinguishes calcite is the chemical makeup of the mineral and a hardness of 3 on the Mohs scale. More than 800 types of calcite have been classified! Marble and limestone are mostly made of calcite; it is extremely common.

Rock Study

Isn't it amazing that these are all of the same family? Each one is incredibly unique and beautiful. Their colors, shapes, crystal structures, and environments are vastly different, yet they all share a chemical compound that ties them together. What is your favorite thing about each one?

FACTS

TYPE: Mineral

MOHS SCALE: 3

LUSTER:
- Vitreous
- Pearly

Wulfenite

Wulfenite is unique because of the tabular shape of its crystals. Wulfenite in its pure form has no color, but nearly all samples range from creamy yellow to orange or brown or even red. The beautiful coloring is due to the presence of chromium.

Rock Study

Wulfenite is named in honor of Franz Xaver von Wulfen. Most minerals are named for their appearance, content, and other characteristics. Look closely at the colors and tabular shape of the crystals. If you were the scientist who discovered it, what would you name it, based on its appearance?

FACTS

TYPE: Mineral

MOHS SCALE: 3

LUSTER:
- Adamantine
- Resinous

QUARTZ

Quartz is the second most abundant mineral in the crust of the earth. There are many, many varieties of quartz, some of which are semi-precious gems, such as amethyst and agate. In its pure form, it is colorless and very hard. It cannot be scratched by steel!

ROCK STUDY

Look at the pictures of quartz crystals. Isn't it incredible that these beautiful crystals grew this way in nature? These were not cut by people. This is the way they formed naturally. Notice the different colors of the quartz. Don't they make the crystals interesting?

FACTS

TYPE: Mineral

MOHS SCALE: 7

LUSTER: Vitreous

FUN FACT: Quartz crystals are piezoelectric, and they are very useful in electronics such as telescopes, clocks, and camera lenses.

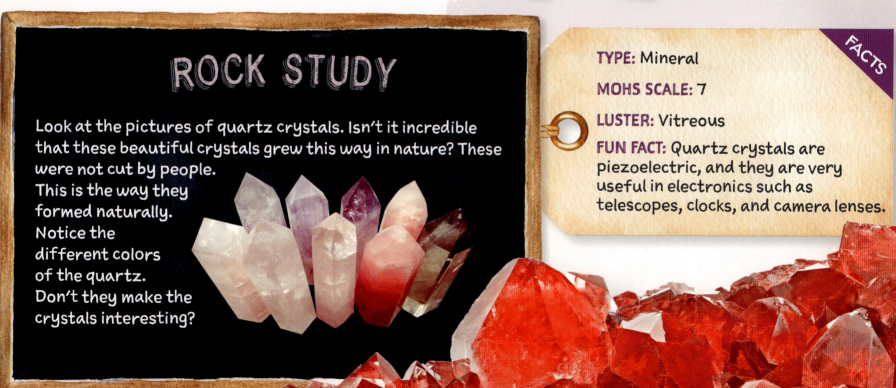

Ruby

A red variety of corundum is a mineral widely known as a ruby. Rubies are one of the hardest minerals on the planet, next to diamonds and moissanite. They range in color from pink to a really deep blood red. The most valuable ruby, called the Sunrise Ruby, sold for over $30 million in 2015!

TYPE: Mineral

MOHS SCALE: 9

LUSTER:
- Subadamantine
- Vitreous

ROCK STUDY

Look at the two pictures below. Can you tell which picture shows rubies and which shows fruit? Which fruit looks like an abundant mass of rubies when you peel back the skin? A pomegranate! How do the rubies look similar to the fruit? How do they look different?

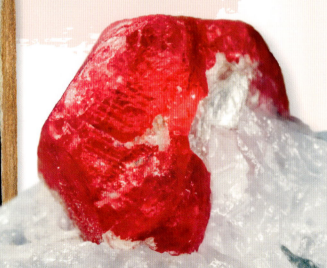

Sapphire

A sapphire is a variety of the mineral corundum. It is most notably blue, but it also occurs in purple, yellow, green, and orange colors. The other variety of a corundum is red, and it is called a ruby. Sapphires and rubies have been found growing next to each other in the same mine, except the rubies grew on marble, while the sapphires grew on granite-like mineral rocks.

Rock Study

Look at the variety in the color of sapphires. The orange one, called a padparadscha sapphire, is a mix between orange and pink and is extremely rare. Notice its clarity (no specks inside it). The higher the clarity, the more valuable the gem. Which of these sapphires below have visible inclusions (mineral specks)? It may be difficult to tell without turning them in your hands!

FACTS

TYPE: Mineral

MOHS SCALE: 9

LUSTER:
- Subadamantine
- Vitreous

Diamond

FACTS

TYPE: Mineral

MOHS SCALE: 10

LUSTER:
- Adamantine

Diamonds are among the most valuable materials on the earth due to their rarity, hardness, and brilliance. But, in space, diamonds are plentiful! Three percent of all carbon found in meteorites is in the form of diamonds. Scientists believe that some planets in our solar system (Jupiter, Neptune, and Saturn) may actually "rain" diamonds!

Rock Study

Look at the close-up diamond. Doesn't it look like a kaleidoscope? Notice all the colors reflected inside. Imagine turning the diamond. How do you think it would look as you turned it? Do you see all the colors of the rainbow? What colors are missing? Which color is most plentiful?

JADE

Jade refers to two different minerals that are very similar in hardness and appearance: jadeite and nephrite. Jadeite is slightly harder and more valuable. The more iron present in jade, the deeper the green. Jade can also be found in different colors: white, pink, lavender, and even black. The color depends upon the elements present, such as iron and chromium.

Black Jade

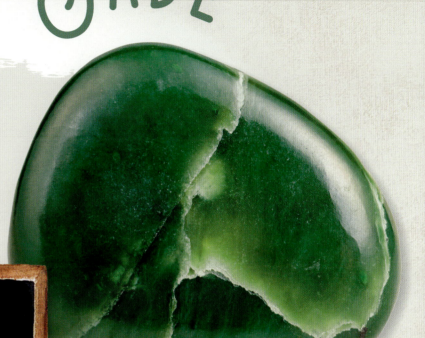

ROCK STUDY

Jade can be quite valuable. The more transparent, translucent, and deep green, the more valuable it is. There are, however, many imitations made from glass or less valuable stones. Real jade is noticeably cold to the touch and quite heavy—more so than glass or even quartz. Imagine what it would feel like to touch the cold and heavy jade.

FACTS

TYPE: Mineral

MOHS SCALE: 6.5–7

LUSTER:
- Vitreous

Charoite

This purple stone was first discovered in the 1940s and was called "Lilac Stone." It is now named after the Chara River in Siberia, Russia, which is the only region in which charoite has ever been found. As rare as it is, it can be found in large chunks and is rather inexpensive. You can buy a stone for a couple of dollars.

FACTS

TYPE: Metamorphic

MOHS SCALE: 5–6

LUSTER:
- Dull
- Pearly
- Vitreous

Rock Study

Charoite has unique swirling patterns due to the inclusion of other minerals and the way its crystals are formed. Sometimes they almost look like pictures of landscapes, oceans, or objects. Find a purple crayon (or multiple shades of purple), as well as 1–2 other colors, and create a swirling pattern of your own on a piece of paper.

Obsidian

Obsidian is a volcanic glass that is a result of lava cooling quickly from contact with a lake, ocean, or cool air. Obsidian can be made into extremely sharp blades. Even the sharpest metal blade, when viewed under a microscope, has jagged edges, but obsidian remains smooth.

Rock Study

Snowflake obsidian gets its name because of the inclusion of cristobalite crystals that form inside the obsidian, creating a splotchy look. Notice the stark contrast between the black obsidian and white crystals. How many snowflake shapes do you see on the stone?

FACTS

TYPE: Igneous

MOHS SCALE: 5–6

LUSTER:
- Vitreous

Emerald

Emeralds are one variety of the beryl family. The distinguishing factor for emeralds is the green color, which is a result of the presence of chromium or vanadium. They range in color from yellow-green to blue-green, but the deeper the green color and the clearer the stone, the more valuable it is.

FACTS

TYPE: Mineral

MOHS SCALE: 7.5–8

LUSTER:
- Vitreous

ROCK STUDY

Most emeralds have inclusions (specks, lines, etc.) and fissures, or tiny cracks. Emeralds without inclusions and fissures are very rare and more valuable than other emeralds. Look closely at the pictures. Which ones appear the most clear? Which has the fewest inclusions and fissures? Which do you think would be most valuable?

Bloodstone

Bloodstone (also called heliotrope) gets its name from the flecks of red in the green opaque stone, due to hematite (iron oxide) inclusions. The name hematite comes from the Greek word for blood. Bloodstone has traditionally been used for jewelry, figurines, and small carvings.

Rock Study

Not all bloodstones have the red inclusions of hematite. Look at this picture with eleven bloodstones. How many of them appear to have the red specks? How many do not? Look at the heliotropes with no red inclusions. Notice the difference in color, from light to dark green. Which do you like best?

FACTS

TYPE: Igneous

MOHS SCALE: 6.5–7

LUSTER:
- Vitreous to waxy

Jasper

Jasper is made of tiny quartz crystals. There are numerous varieties, and they are found all over the world. Because it is a harder stone, it is often used in jewelry, though not usually faceted because it is opaque. Jasper is often carved and made into decorative items, such as this lovely vase. Bloodstone is a variety of jasper.

Rock Study

Isn't it amazing how unique each piece of jasper is? Look at each stone. Notice the many colors made up of mineral impurities. Note how some are splotchy, some have bands, others seem to flow like a winding river, and some have speckles. What story do you think each one is trying to tell?

FACTS

TYPE: Mineral

MOHS SCALE: 6.5–7

LUSTER:
- Waxy
- Dull
- Vitreous

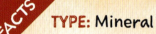

22

Magnetite

Particles of magnetite commonly exist in metamorphic and igneous rocks. It is mined today as iron ore. Magnetite has been found in the brains of birds, bees, and even humans. It is believed that the magnetite in brains works with the magnetic field of the earth, helping a variety of species, including humans, with their sense of direction.

Rock Study

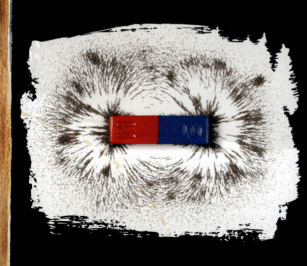

Magnetite is attracted to magnets and can also be magnetized to be a permanent magnet. If you have a magnet and access to sand, rub the magnet through the sand. It should attract tiny bits of magnetite that stick to it, similar to the picture here.

FACTS

TYPE: Igneous

MOHS SCALE: 5.5–6.5

LUSTER:
- Metallic to submetallic

Fluorite

Fluorite grows in the shape of cubic crystals, and though naturally colorless, it is usually found in an assortment of colors, depending on nearby minerals. Fluorite is fluorescent, meaning it glows under a black light. The word "fluorescent" comes from this glowing mineral. Fluorite found in different places might glow different colors!

FACTS
TYPE: Mineral
MOHS SCALE: 4
LUSTER:
- Vitreous

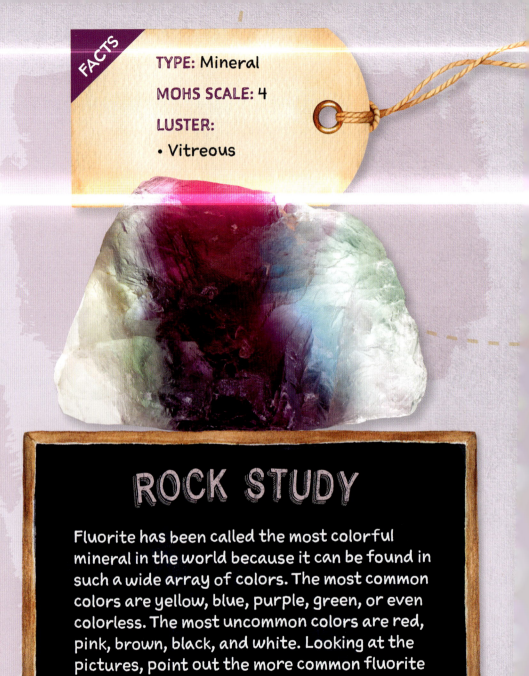

ROCK STUDY

Fluorite has been called the most colorful mineral in the world because it can be found in such a wide array of colors. The most common colors are yellow, blue, purple, green, or even colorless. The most uncommon colors are red, pink, brown, black, and white. Looking at the pictures, point out the more common fluorite stones, and then find the less common ones.

Gypsum

Have you ever used plaster of paris or sidewalk chalk? They are made from a very soft mineral called gypsum.

FACTS

TYPE: Mineral

MOHS SCALE: 1.5–2

LUSTER:
- Vitreous to silky
- Waxy
- Pearly

ROCK STUDY

Look at the pictures of gypsum. Think about the feel of a soft piece of chalk in your hand. Look at the gypsum sand dunes found in New Mexico. How much chalk do you think those dunes would make? Now look at the giant gypsum crystals found in a cave in Mexico. Do you think they would hold you if you climbed on them, or would they crumble?

This highly useful mineral is also used in medical casts for broken bones, drywall for construction, cement, and fertilizer.

Mica

Mica is a highly flexible group of minerals. You can bend thin sheets of it, and it will go back to its normal shape. Because of the structure of its crystals, you can break mica apart into thin sheets. It is also fairly soft; you can scratch it with only your fingernail. It is used for many things, from electric wiring and laser devices to joint compound for drywall. It also creates the sparkle in makeup.

FACTS

TYPE: Mineral

MOHS SCALE: 2.5–4

LUSTER:
- Vitreous
- Pearly

ROCK STUDY

Look at the pearly shine of the mica. Picture yourself sliding your fingernail under the edge of the mica and gently peeling apart a thin sheet from the rock. You can almost see through it. Imagine the sheet bending back and forth in your fingers. It may make a crinkle sound, but it won't break if you are gentle!

Tiger's Eye

ROCK STUDY

Take a close look at the tiger's eye stones. If you were to create a watercolor tiger's eye, what color would you start with? Choose one of the stones pictured here to paint. You can paint it, or imagine painting it, starting first with one of the lightest shades. Then add the darker stripes. Add some of the darker specks. Notice how the stripes go one way while the tiny fibers go perpendicular to the stripes.

Tiger's eye is a type of quartz with iron inclusions. The striped appearance occurs because the semi-precious stone starts out as another fibrous mineral that later turns to quartz. It has a quality called "chatoyance," a visual effect (also called "cat's eye effect") created when there are fibrous minerals included in the stone.

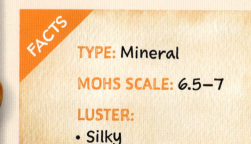

FACTS

TYPE: Mineral

MOHS SCALE: 6.5–7

LUSTER:
- Silky

Turquoise

Because of its striking blue to blue-green color, turquoise was a quite valuable gemstone for years. Then the market was flooded with man-made versions, as is common with opaque stones. Gemologists can often tell the fake from the genuine by using magnification.

FACTS

TYPE: Mineral

MOHS SCALE: 5–6

LUSTER:
- Waxy to subvitreous

Rock Study

Can you tell which stones in the pictures are real and which are fake? There are two of each. Can you see how easily someone could be fooled? Notice how even the mineral veins are imitated. One of them has pyrite inclusions. Can you pick it out? That one is a genuine stone.

Numbers 2 & 3 are real turquoise

28

Malachite

Malachite's bands of light and dark green make it beautifully unique. It was named after the mallow plant, because it resembles the plant's green leaves. From ancient times until about 1800, crushed malachite was widely used to make green paint. The ancient Egyptians also ground it into a powder for use in makeup.

Rock Study

As you look at the malachite stones, can you see the difference in luster of each band of color? Luster refers to how light interacts with the mineral. Some bands are more opaque. Some look glassy (vitreous), others waxy (as if coated in wax). A dull luster means it is porous or rough. Point out the various lusters.

FACTS

TYPE: Mineral

MOHS SCALE: 3.5–4.5

LUSTER:
- Adamantine
- Vitreous
- Silky
- Dull

Peridot

Also called chrysolite, peridot is always some shade of green—from yellowish-green, to olive green, to lime green, and even brown-green. It is one of two gems not formed in the earth's crust but in the earth's mantle; the other is diamond. It is brought to the earth's surface by volcanic activity and is found in igneous rocks.

FACTS

TYPE: Mineral

MOHS SCALE: 6.5–7

LUSTER:
- Vitreous to oily

Rock Study

When a stone is faceted, it is cut to have many different faces. This increases its shine and brilliance. The stone is also polished. Notice how the faceted stones seem to have many shades of green because of the way the stones' surfaces reflect the light.

Legrandite

Legrandite is a rare gem desirable for its distinctive yellow color. The most spectacular specimens have been found in Mexico, where the mineral was first described by a Belgian mining engineer named Louis C.A. Legrand. Carvers of this stone must take great care because it contains arsenic, a toxic mineral.

FACTS

TYPE: Mineral

MOHS SCALE: 4.5–5

LUSTER:
- Vitreous

ROCK STUDY

Observe that two of the specimens pictured have a glassy (vitreous) luster. They appear to be larger than the other two, but they must have been photographed close-up because legrandite tends to become more opaque the larger it is.

ANGLESITE

This lustrous mineral is named for the Welsh island of Anglesey, where the mineral was first discovered.

TYPE: Mineral

MOHS SCALE: 2.5–3

LUSTER:
- Adamantine
- Vitreous
- Resinous

ROCK STUDY

Notice the hardness of this stone. Do you think it would be a good choice for making jewelry? It is sometimes faceted and placed in pendants and earrings, but it is not hard enough to be durable on a ring. Its surface is too easily scratched.

Notice the brilliant shine of its delicate crystals. Can you see how it grows with many shimmering angles and surfaces? Its colors range from colorless to white or gray, yellow, orange, brown, and even bluish or pale green.

FELDSPAR

FACTS

TYPE: Mineral
MOHS SCALE: 6–6.5
LUSTER:
• Vitreous

Feldspar is the most abundant rock mineral on the earth, making up over 50% of the earth's crust. This mineral is found in igneous, metamorphic, and sedimentary rocks. About 450,000 metric tons of feldspar are produced each year just in the US. Some countries produce more! It has numerous uses: in glassmaking and ceramics (like tile), and as a filler for paint, plastics, and rubber.

ROCK STUDY

Feldspar can be white, gray, pink, reddish, brown, and green. There are several varieties. Pictured here are amazonite, moonstone, sunstone, and raw feldspar. Can you match the name to the stone? Look at the unique details of each one. How do the stones resemble their names?

1

2

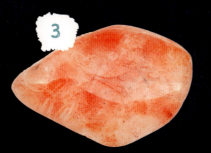
3

4

1. Moonstone 2. Raw feldspar 3. Sunstone 4. Amazonite

Labradorite

This colorful stone is a variety of feldspar. Its unique sheen, called labradorization, occurs when light reflects off the stone in one direction. As you tilt the stone in different directions, the sheen comes and goes. It is named for the province where it was first identified: Labrador, Canada.

FACTS

TYPE: Mineral

MOHS SCALE: 6–6.5

LUSTER:
- Vitreous
- Pearly

ROCK STUDY

Have you ever seen or heard of the northern lights, the lights that dance and flicker across the dark night sky? Labradorite is reminiscent of that phenomenon, because colors flicker across an otherwise dull stone. The stones that show multiple colors at once, such as these, are especially unique.

Graphite

Graphite occurs in metamorphic rocks, igneous rocks, and also in meteorites. If the heat and pressure are high enough, graphite will turn into a diamond! The "lead" in pencils is really graphite. Since it is also a conductor of electricity and heat, it is used in electronics, solar panels, and batteries.

Rock Study

People sometimes confuse graphite with coal. Have you ever done a coal sketch, rather than a pencil sketch? Graphite has a crystalline structure that coal does not, so it is more shiny, and the stone has more structure. Use a pencil to make a dark scribble on a paper. Can you see the sheen that it has? One of these pictures is of coal. Can you pick it out?

2. Coal

FACTS

TYPE: Mineral

MOHS SCALE: 1–2

LUSTER:
- Earthy
- Metallic

RHODOCHROSITE

The most spectacular rhodochrosites are found in Colorado, ranging from deep red to light pink hues. Because they are relatively soft, they can be difficult to cut, but there are cut stones as large as 60 carats! Most faceted stones, however, are 5 carats or less.

FACTS

TYPE: Mineral

MOHS SCALE: 3.5–4

LUSTER:
- Vitreous
- Pearly

ROCK STUDY

As you compare the photos, notice colors, shapes, patterns, and designs. What similarities do you see between them? What differences do you see? What fruits do they resemble? Which are vitreous? Pearly? Both?

Amethyst

Amethyst is a type of quartz that ranges in color from a pale lavender to a dark reddish purple. Extremely beautiful and hard, it was once among the most valuable gemstones (along with diamonds, rubies, emeralds, and sapphires). But since large deposits of them have been found, especially in Brazil, their value has decreased significantly.

Rock Study

Observe the different shades of purple amethyst. How would you describe each color to someone? Would you compare them to something, such as a plum? Describe not only the color but also the shapes or sizes of the crystals or gemstones.

FACTS

TYPE: Mineral

MOHS SCALE: 7

LUSTER:
- Vitreous

Flint

Flint is the sedimentary rock form of quartz, and therefore it is very hard. It will spark when struck with steel and can be used to start a fire, with proper kindling. Flint chips create sharp edges, which made them useful in ancient times for making knives, spears, arrows, and other tools.

FACTS

TYPE: Sedimentary

MOHS SCALE: 7

LUSTER:
- Vitreous

Rock Study

Flint is colored light brown, black, moderate to dark gray, reddish brown, or a grayish white. Which of those colors do you see in the pictures? Which has the most of one color?

Ametrine

FACTS
TYPE: Mineral
MOHS SCALE: 7
LUSTER:
- Vitreous

Ametrine is quartz that is banded with purple and yellow, a combination of amethyst and citrine (a yellow quartz). Ametrine is found in only one place in Bolivia. No other colors of ametrine exist.

ROCK STUDY

We know that ametrine is yellow and purple, but it is much more complex than two simple colors. Look at each piece pictured. Notice the many shades of purple in the faceted stone. Notice the pink and tones of yellow. In the other stones, notice all the inclusions and additional "colors" that the light reflections create.

Spessartine Garnet

FACTS

TYPE: Mineral

MOHS SCALE: 7–7.5

LUSTER:
- Vitreous

ROCK STUDY

Have you ever put a strong flashlight behind your hand to watch the front of your hand glow? If you hold the dark reddish spessartine up to the light, the light shines through, making the stone glow orange. How would you put these stones in order from lightest to darkest?

Rare and valuable, spessartine garnets range in color from light mandarin to a deep reddish-orange to brownish and can be found in several countries around the world. The gems are named after the Spessart Mountains in Bavaria, Germany, where they were first found. The rich mandarin colors fetch the highest prices. Violet-red spessartines have been discovered in Colorado and Maine.

Azurite

Prized for its deep blue crystals, this gem is commonly made into cabochons and decorative carvings. It usually has green malachite inclusions, which create a striking contrast. Both azurite and malachite get their coloring from copper, but the copper creates different colors in different minerals.

FACTS

TYPE: Mineral

MOHS SCALE: 3.5–4

LUSTER:
- Vitreous

Rock Study

This looks like a river running through a jungle! Let your finger float along the blue river, avoiding the trees.

Would you like to wear this necklace? Can you imagine the cold, weighty stones?

What does this one look like to you?

This stone resembles space and looks like a spaceship! Blast off!

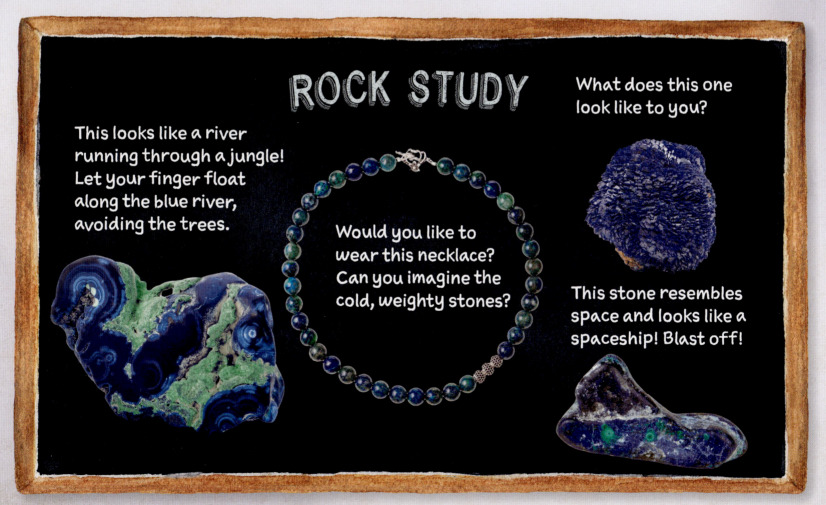

Unakite

FACTS

TYPE: Metamorphic

MOHS SCALE: 6–7

LUSTER:
- Vitreous
- Glossy

Unakite is a metamorphic rock that was previously granite, but it was changed by hydrothermal movement. During that change, granite was replaced with a pink feldspar and another mineral called green epidote.

ROCK STUDY

Choose one of these unakite stones that looks fun to paint. You can either paint it or imagine painting it. What color will you start with—green or pink? You may need to mix some colors to get just the right hues. After painting the green and pink, add the darker veins of gray, brown, or black. Some have speckles of white. Have fun with your real or imaginary painting!

Morganite

Named after J.P. Morgan, a famous US financier and banker, morganite is a semi-precious gem of pink, salmon, or purplish pink. High on the hardness scale, morganite could make a beautiful ring and would last a lifetime or more. It is closely related to the gems aquamarine and emerald, as they are all beryls.

Rock Study

Notice the difference in clarity of the faceted gems compared to the other stones and crystals. The faceted gems have a much higher clarity, which makes them more valuable. The higher-clarity gems have been faceted to enhance their shine. Which stone would you like to have?

FACTS

TYPE: Mineral

MOHS SCALE: 7.5–8

LUSTER:
- Vitreous